AMOEBA

(uh·mee·buh)

Meet these two amazing amoebas! An amoeba is a tiny, shape-shifting acrobat. Amoebas can change their form to squeeze through tight spaces. Their blob-like bodies engulf tiny particles like a hug. Despite their small size, amoebas play a big role in the world, adding to the balance of life in ponds, lakes, and even the soil beneath our feet.

B ACTERIO-PHAGE

(bak·tee·ree·ow·fayj)

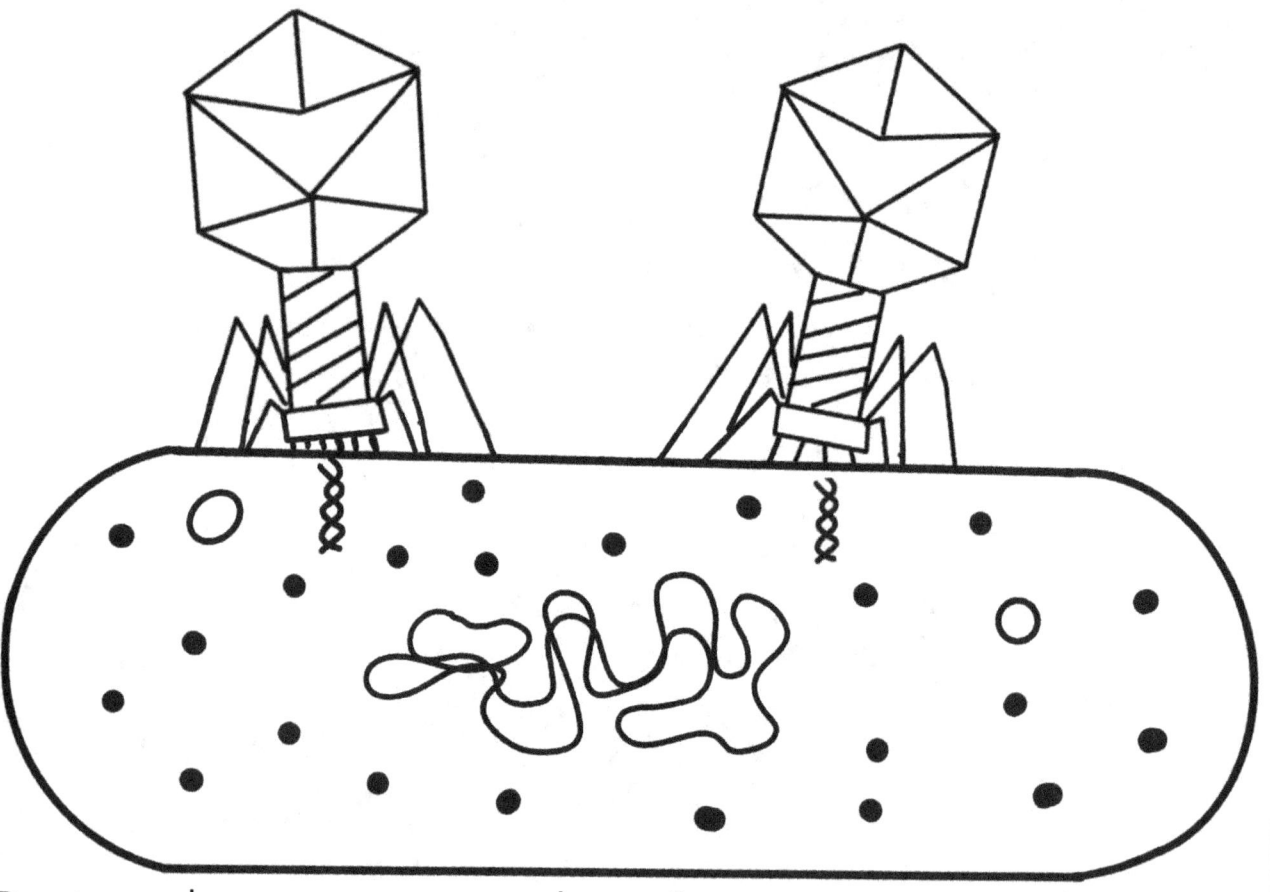

Bacteriophages are viruses that infect certain types of bacteria. They attach to the bacteria, inject their genetic material, and create new phages inside. When enough of these little viruses are made, they burst out of the bacterial home to find more bacteria to infect. One day, scientists hope they can be used to fight infections in humans!

CILIATE

(si·lee·uht)

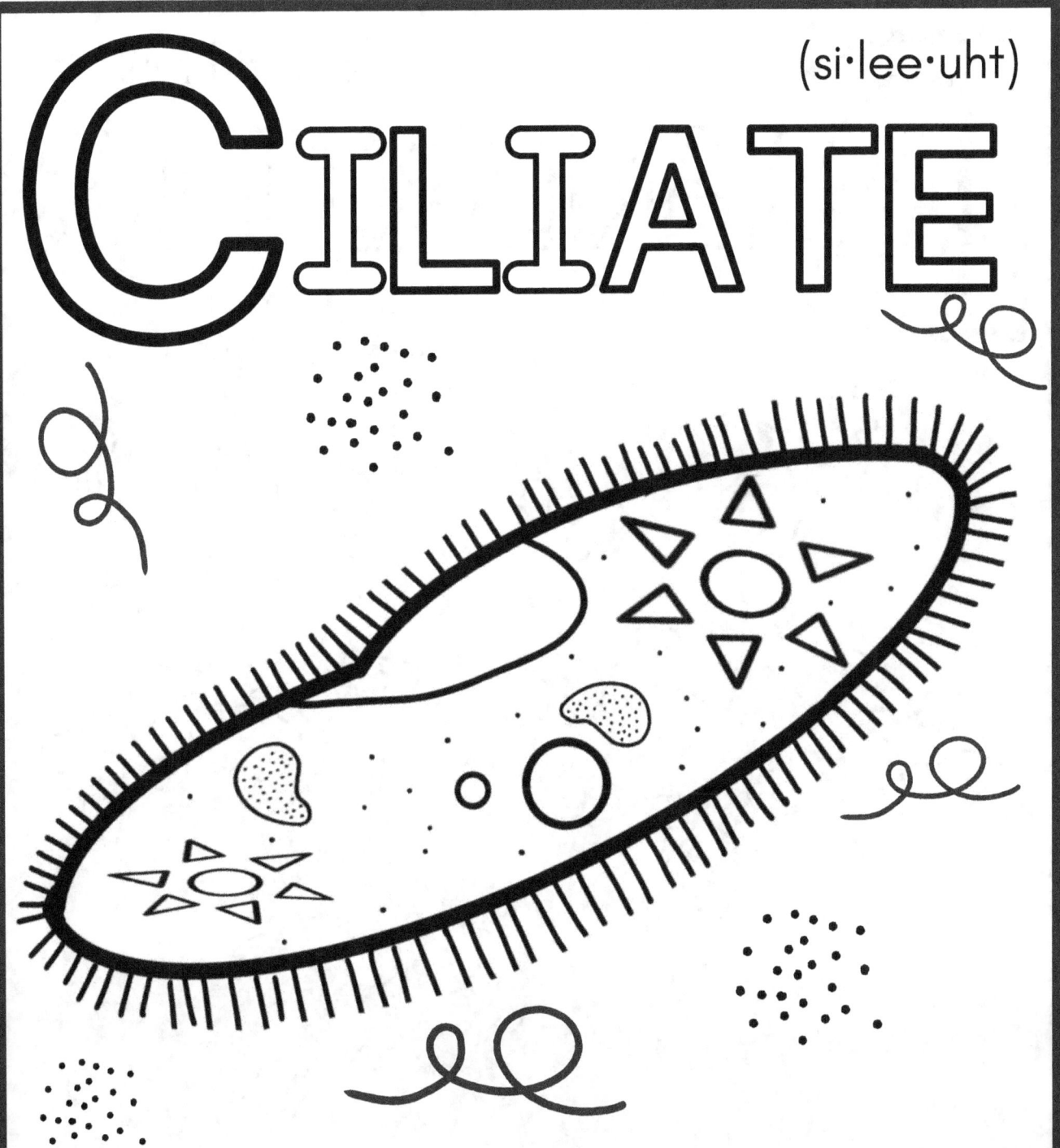

Dive into the world of ciliates! Ciliates are named for their hair-like cilia that surround the outside of their cell. The ciliate you see above is a paramecium, which uses its cilia to glide through the watery world. Ciliates can use their cilia to create currents that sweep food into their mouths, making them excellent hunters. See the paramecium twirling and gliding, capturing microscopic snacks.

Dendritic Cell

(den·dri·tuhk sel)

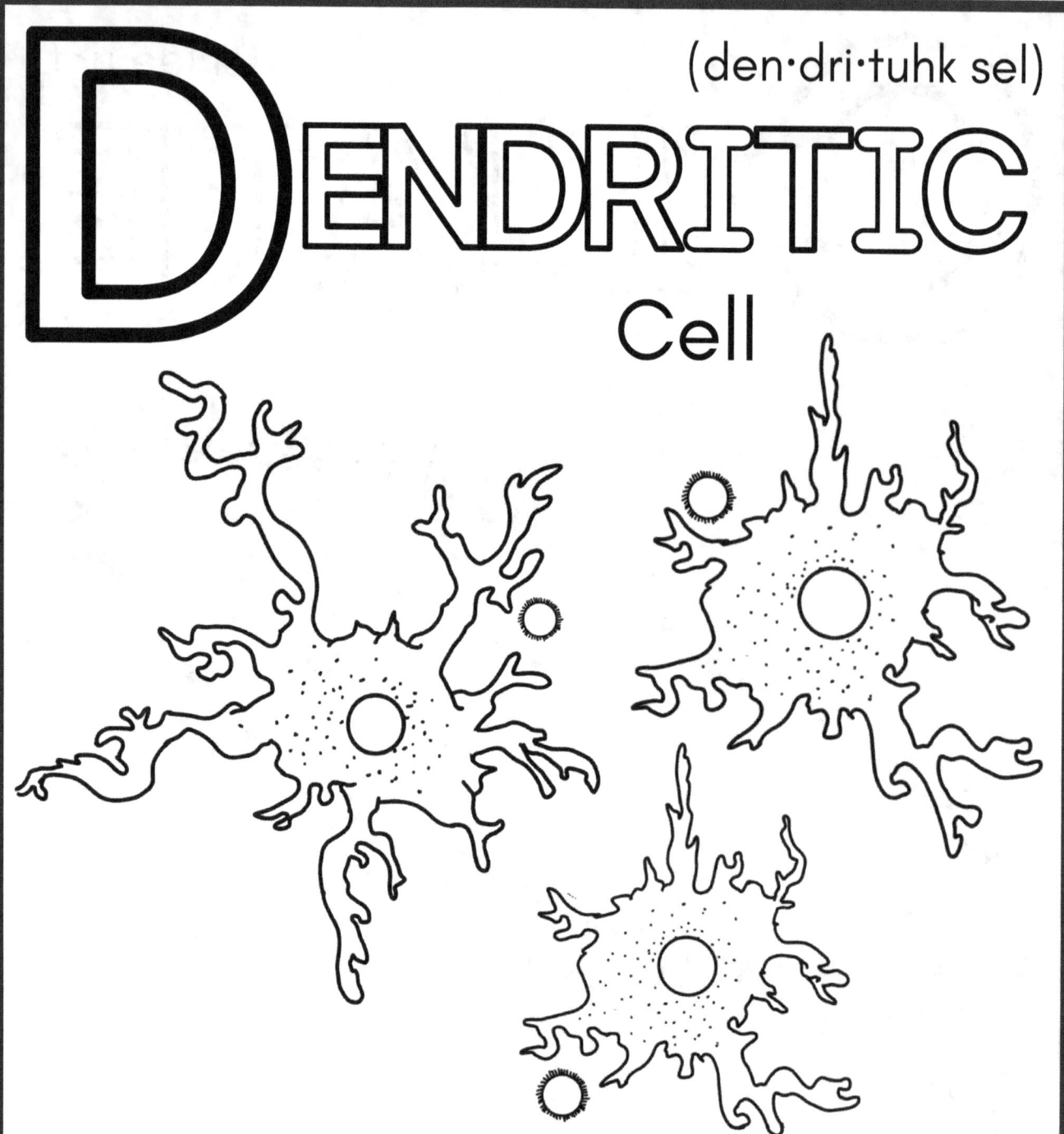

The dendritic cell – an immune superhero with tree-like branches called dendrites. Dendritic cells patrol our body, capturing and presenting invaders, like viruses, to other immune cells. They are the messengers that guide our immune system to protect us from germs. Color the dendritic cells and imagine them leading the immune team in a battle against infections! Stay strong, little defenders!

E UGLENA

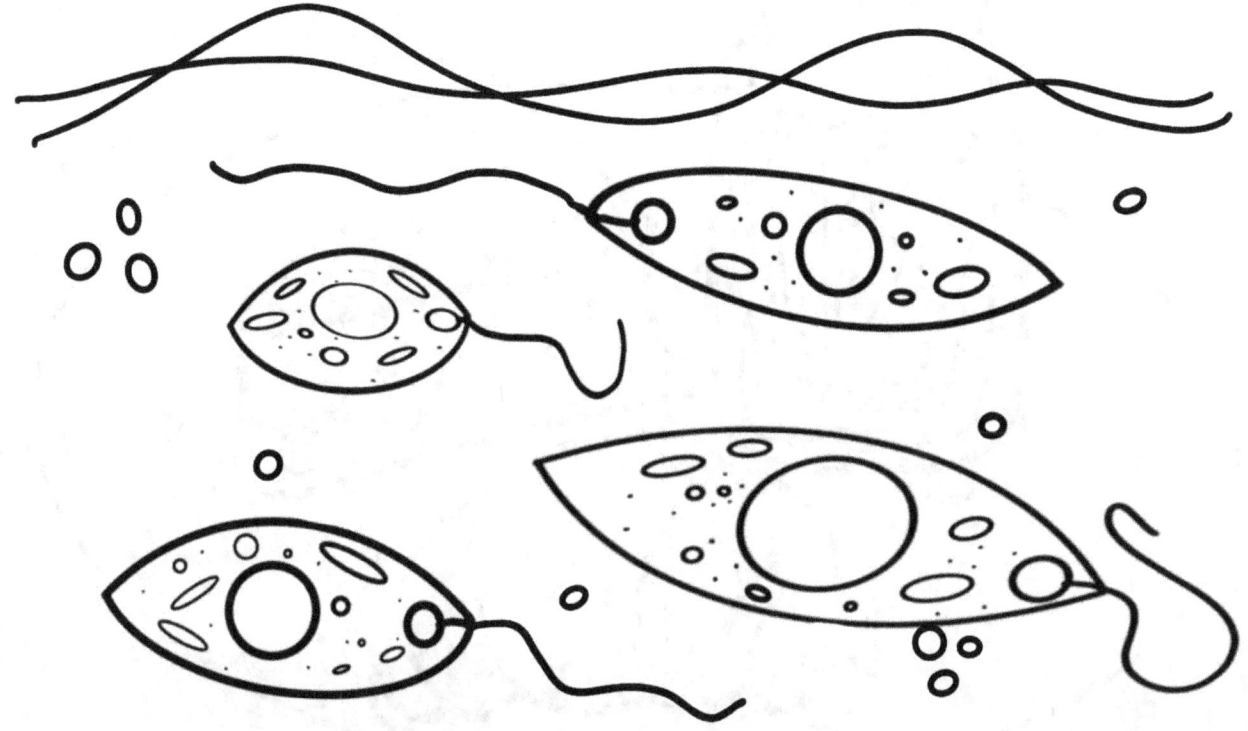

Embark on a journey into the unseen world and discover the fascinating euglena! Picture a tiny creature with a whip-like tail, called a flagellum, swiftly navigating freshwater habitats like ponds and lakes. The euglena is not just a swimmer but it's also like a miniature plant, using sunlight and chlorophyll to create its food. Color the euglena in shades of green.

Fermentation

(fur·muhn·tay·shn)

Do you like foods like cheese, pickles, or bread? A process called fermentation makes them possible! Fermentation is a way microorganisms help us make these foods. Bacteria and yeasts use sugar to create the flavors in cheese, the sour taste in pickles, and the gas that makes bread rise. Microorganisms are very helpful in making many different foods we eat each day.

(gram stayn)

GRAM Stain

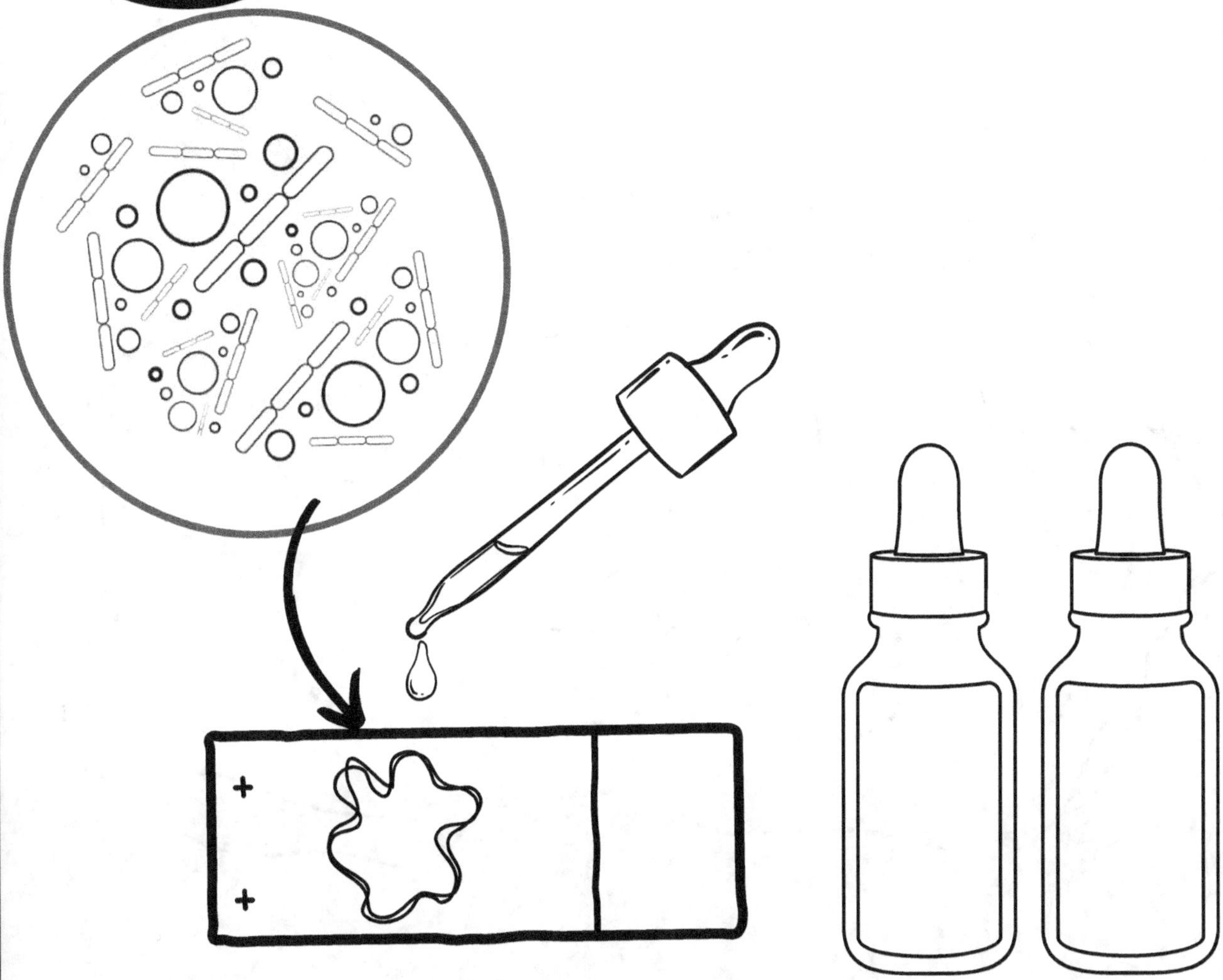

One way to distinguish bacteria is by their shape. Scientists also classify bacteria through a process called gram staining, revealing the type of cell wall they have. Cell walls make up the outside of the bacteria. Gram-positive cells turn purple, while Gram-negative cells turn pink. The cocci (spheres) are positive, and the bacilli (rods) are negative. Use your coloring skills to match the correct colors!

H ALOPHILE

In the salty corners of our Earth, there's a special group of microorganisms known as halophiles. These tiny archaea, bacteria, algae, and fungi thrive in places where most others would find it too salty to handle. They can be found in salt marshes, lakes, brine pools deep in the ocean, and salt mines.

I NOCULATION

(i·naa·kyuh·lay·shn)

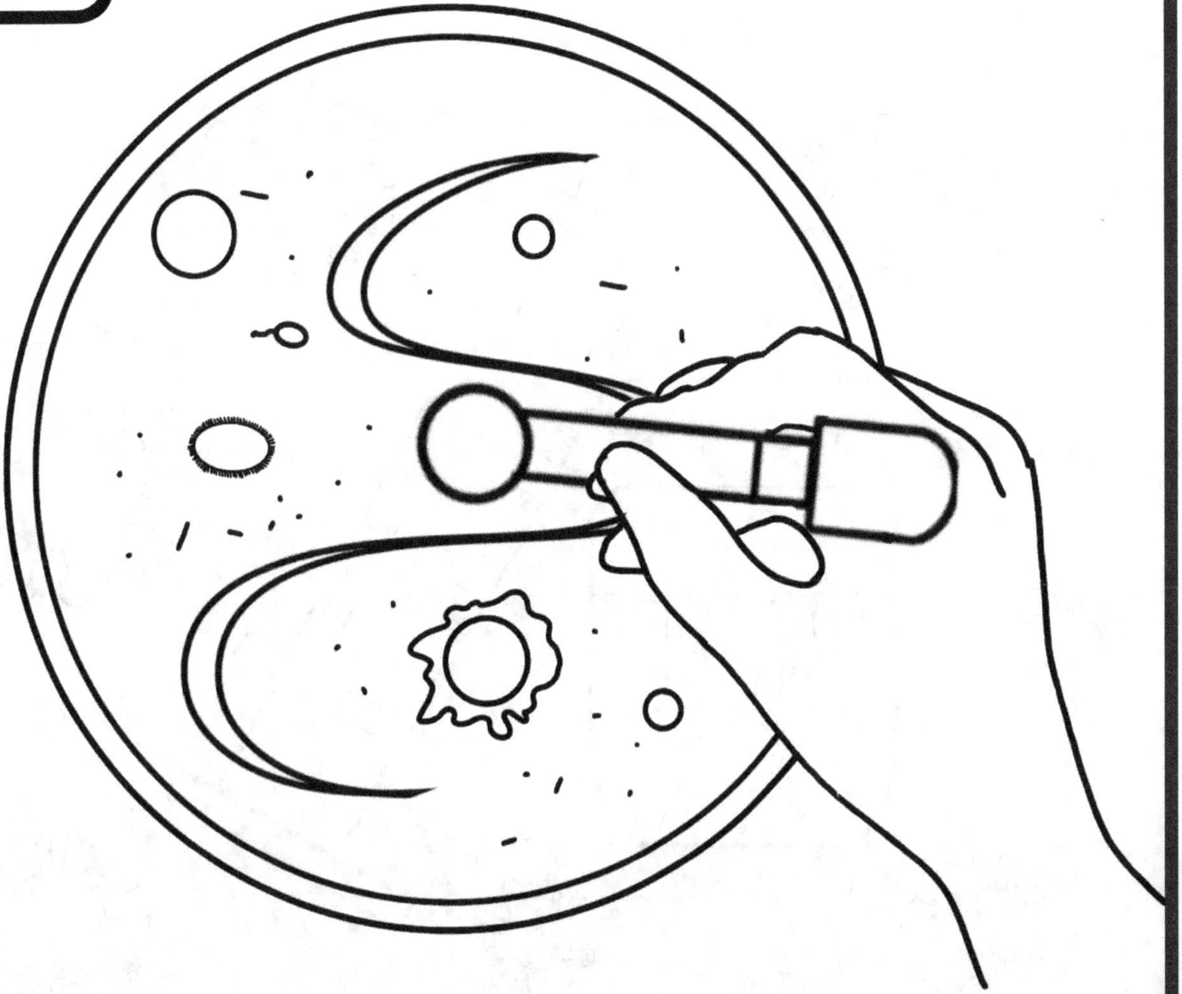

Inoculation means putting a tiny bit of microbes, like bacteria, into a small dish filled with a substance called agar. Agar is like a jelly with nutrients to help the bacteria grow. We use a metal loop to move the bacteria. First, we heat the loop to make it clean. Then, we pick up a little bit of the bacteria and place them in the agar. Here they will grow so scientists can study them.

JENNER

Edward Jenner, an English doctor, pioneered the concept of vaccines. He saw that milkmaids who had been exposed to cowpox seemed to not get smallpox. He did an experiment and successfully gave a boy the cowpox virus, but later gave him smallpox and he didn't get sick. This groundbreaking work laid the foundation for vaccination, offering a powerful tool in the fight against infectious diseases.

K LEBSIELLA

(kleb·zee·eh·luh)

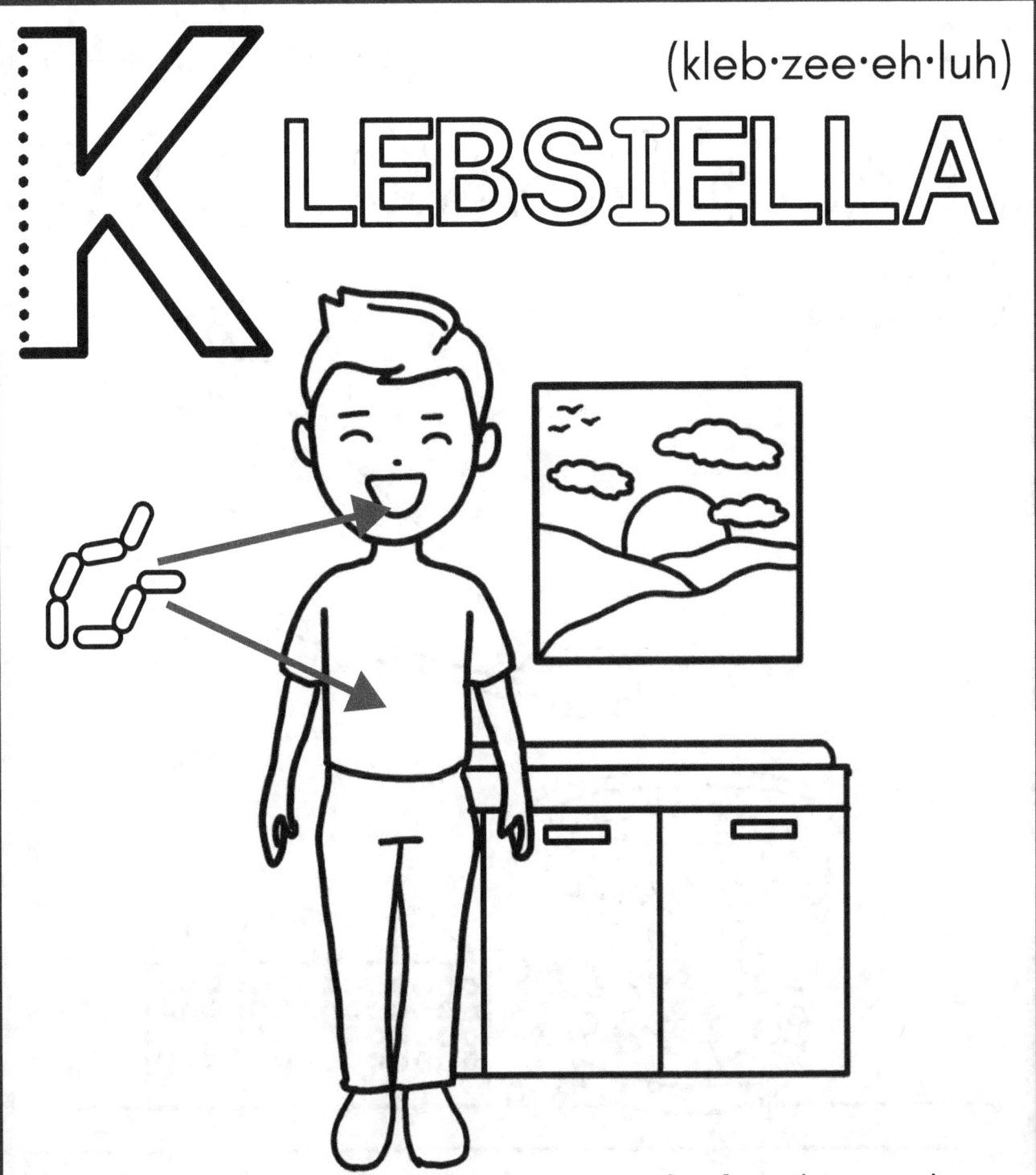

Klebsiella is a family of bacteria that can be found everywhere in nature. It can be found in water, soil, plants, insects, and humans. *Klebsiella* bacteria are typically rod-shaped under a microscope. In humans, the bacteria is found mostly in the nose, mouth, and intestines. Sometimes, when *Klebsiella* grows too much in our bodies, it can make us sick.

LACTOBACILLUS

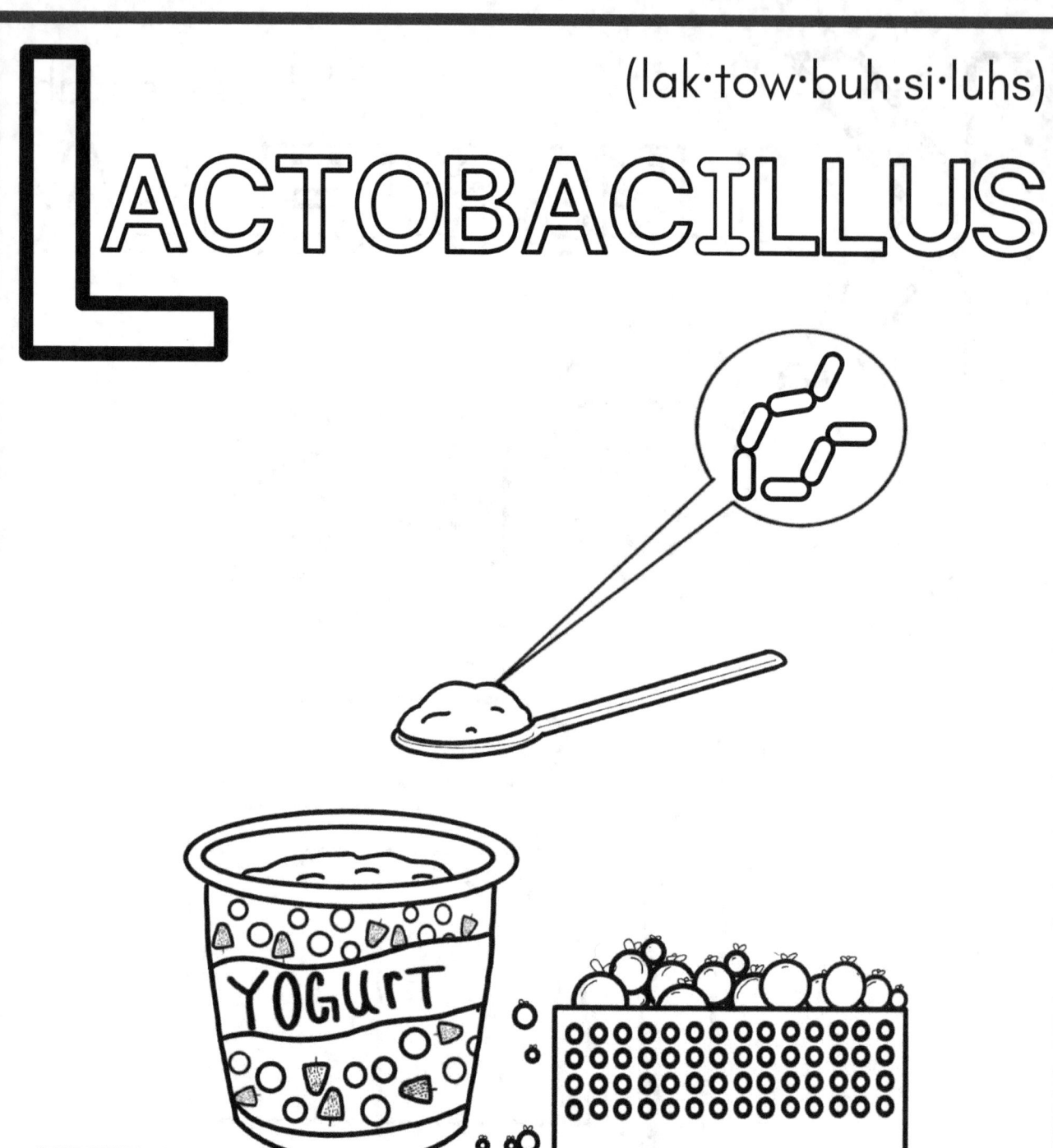

Lactobacillus is a rod-shaped bacteria. It helps make yogurt yummy! These little helpers eat the milk sugars and turn them into lactic acid, giving yogurt its tangy taste. They also make the yogurt thick and creamy. Lactobacilli aren't just found in our yogurt, they can also be found in our bodies. They help our bodies digest food, absorb important nutrients, and fight off bad bacteria!

M ICROSCOPE

(mai·kruh·skowp)

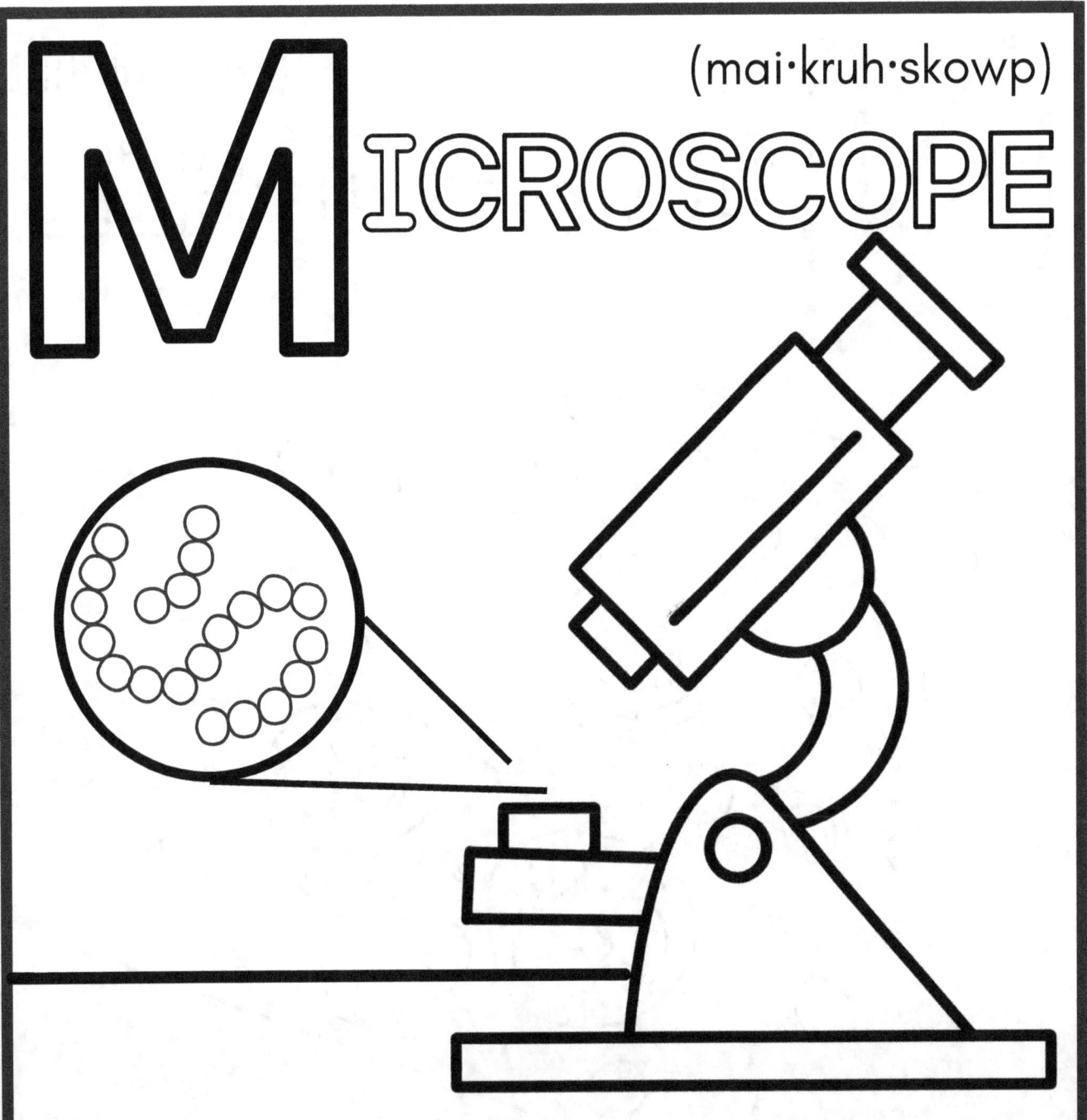

A microscope is a special tool that lets us see very small things, like tiny cells or viruses. It has lenses that make small objects look much bigger! Scientists use microscopes to explore and study the details of the world that are too small to see with just our eyes, like the patterns on a butterfly's wing or the cells that make up plants, animals, and bacteria.

N EUTROPHIL

(noo·truh·fil)

Neutrophils are another important germ-fighting cell in our body. Neutrophils can be found wandering around the bloodstream looking for new bacteria to hunt down and destroy. When it finds a new infection it quickly eats them up and eliminates them. When you see a neutrophil under the microscope, you can see it has a purple, swirly-shaped nucleus.

OUTBREAK

(owt·brayk)

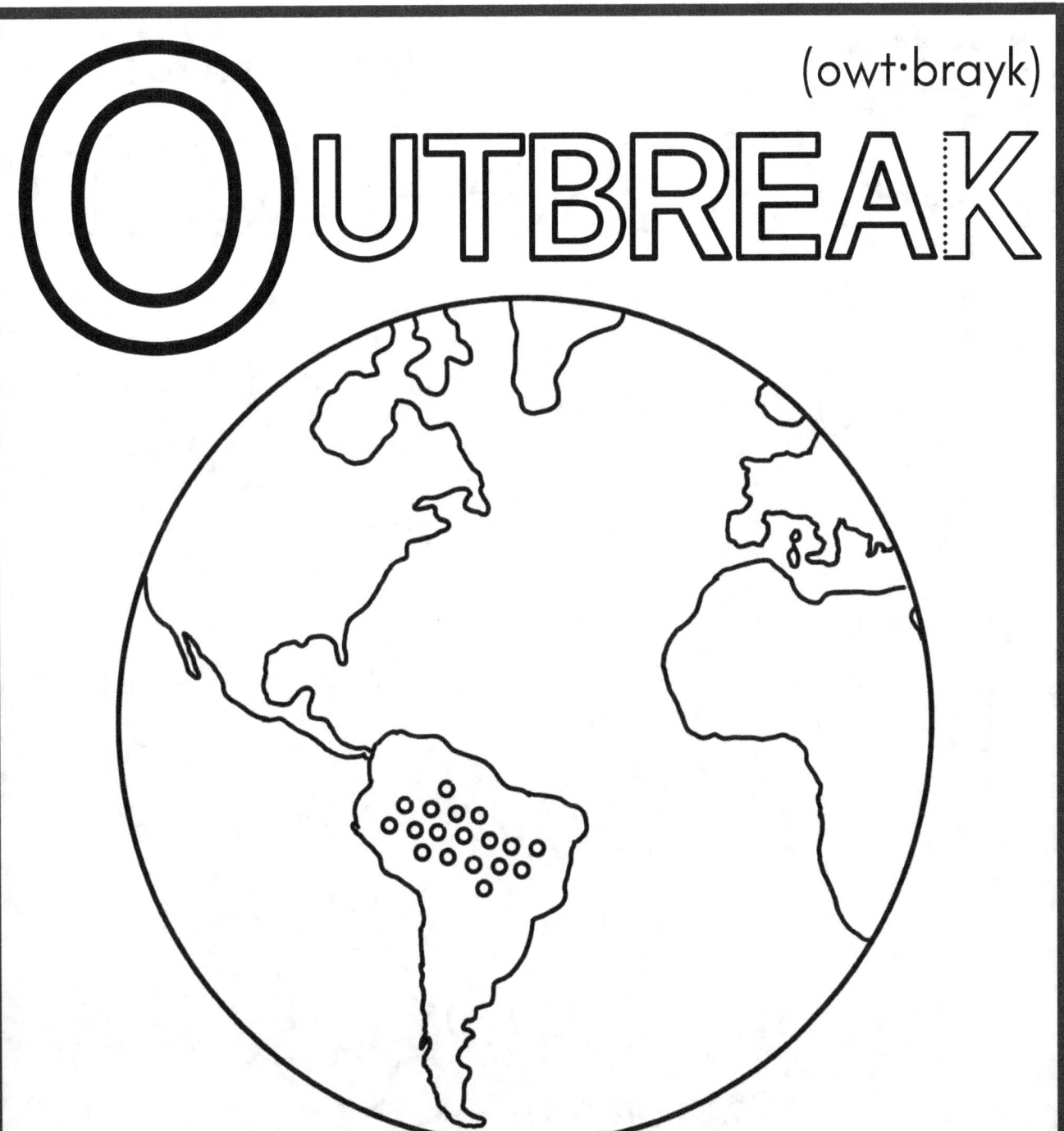

An outbreak, also called an epidemic, in microbiology is when a lot of people in one area get sick from the same germs around the same time. It's like when many people in a school or a neighborhood catch a cold or another illness. Scientists and doctors pay close attention to outbreaks to understand how the germs spread and to find ways to stop them, so fewer people get sick.

P ROKARYOTE

(prow·keh·ree·oht)

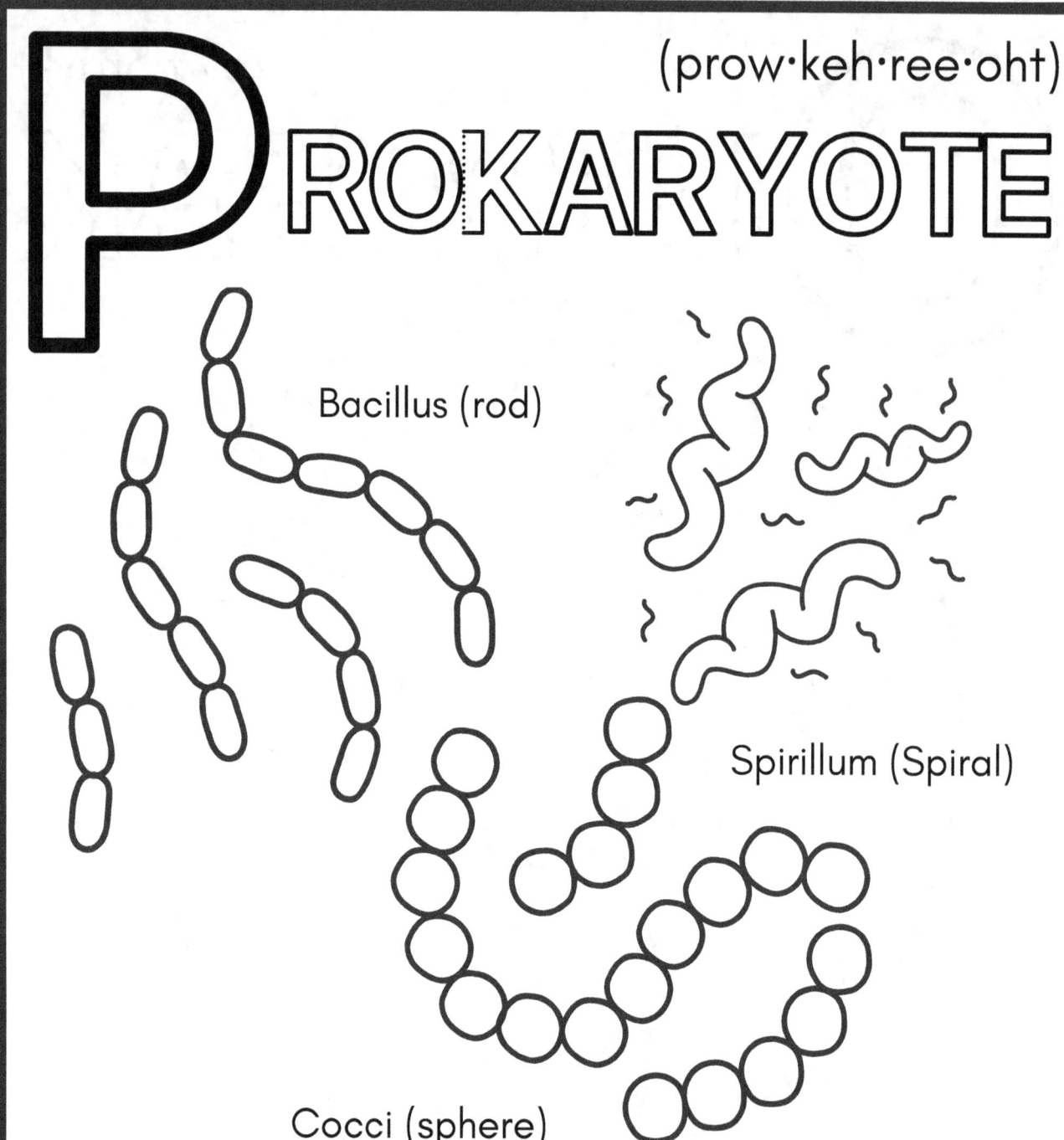

Bacillus (rod)

Spirillum (Spiral)

Cocci (sphere)

Prokaryotes are tiny living organisms that do not have a nucleus and are very simple. Bacteria and archaea are types of prokaryotes. They come in different shapes: spheres, rods, and spirals. Prokaryotes are very important to us and our environment. They help processes like digestion in your gut, while others play crucial roles in helping plants. Some may cause you to be sick, but mostly they can be helpful.

QUORUM

(kwaw·ruhm sen·suhng)

Sensing

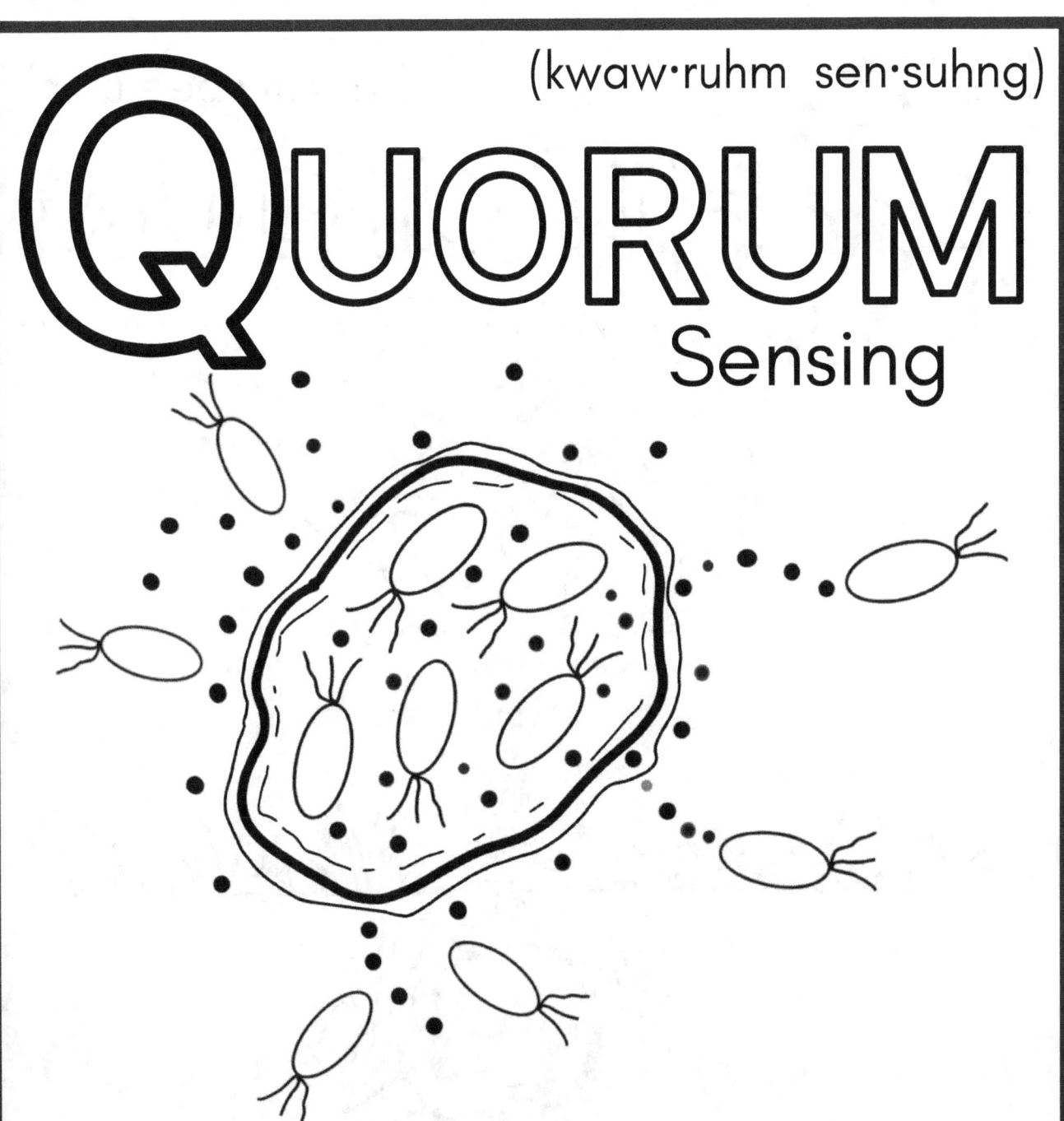

Bacteria have a way of talking to each other, called "quorum sensing." Instead of using words, they release tiny molecules into the environment. When enough of these molecules gather, it signals to the bacteria that there are many friends around. When the bacteria sense that a lot of them are present, they work together. This teamwork can help them form protective layers called biofilms or even launch an infection if they are harmful bacteria.

R HIZOBIUM

(rai·zow·bee·uhm)

Rhizobium is a group of bacteria that are helpful to plants. They form a partnership with some plants, like peas, beans, and clover. These bacteria live in the soil and make a home on the plant's roots, in a bundle called a nodule. Now, what's cool is that rhizobium can take nitrogen from the air and turn it into a form that plants can use to grow big and strong.

SLIME Mold

(slime mowld)

Slime molds are like nature's mystery goo! They live in wet places, like the forest. They like to grow on dead and decaying matter, like logs. Despite their name, slime molds aren't molds at all, they belong to a unique group called protists. Slime molds teach us how cells move and work together, helping scientists learn cool stuff about how living things grow and act.

T

(thur·muh·fai·uhl)
HERMOPHILE

Meet the thermophile, the heat-loving microorganism! These special microorganisms live in super-hot places like hot springs and volcanic dirt. Even though it's scorching for us, thermophiles thrive there. Scientists love studying them because they teach us how life can exist in extreme conditions. Understanding thermophiles helps scientists explore how living things survive in different places, even where it's as hot as an oven!

U BIQUITOUS

(yoo·bi·kwuh·tuhs)

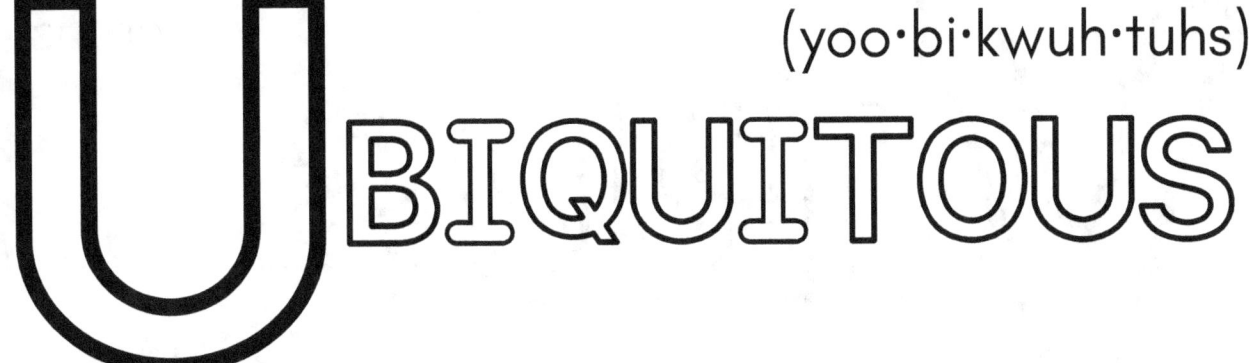

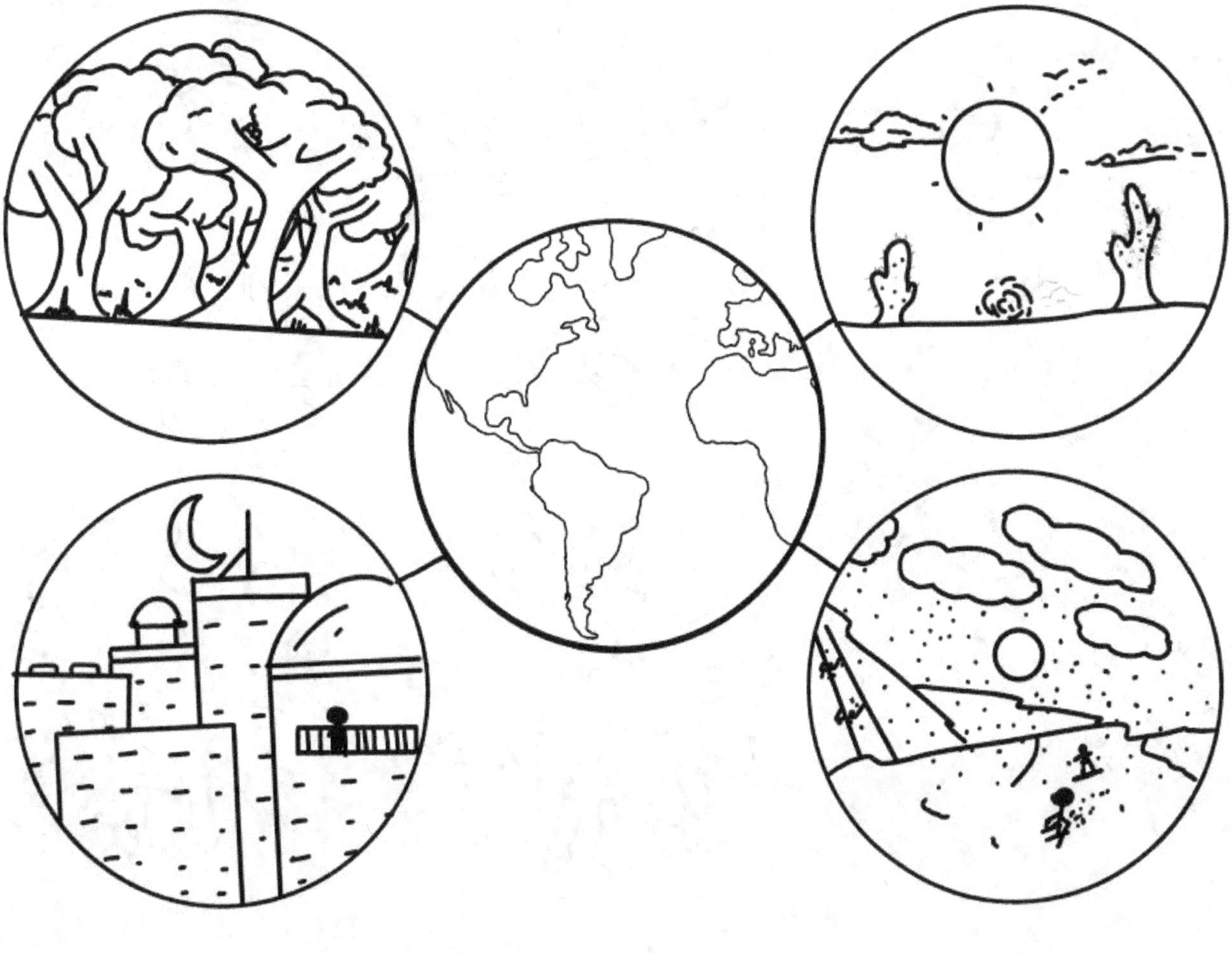

Ubiquitous means it can be found everywhere. Microorganisms are found all around us. They can be where it is super hot, cold, wet, dry, or salty. They can be found in humans, plants, soil, the air, or the ocean. Microbiologists love studying these tiny organisms to understand how they can be everywhere and how they help or harm our world.

VIRUS

(vy·ruhs)

Viruses are super tiny particles that aren't alive like plants or animals but need a host to survive. They're found everywhere, in the air, water, and inside living things. While some viruses can make people, animals, or plants sick, others are crucial for life. They can affect how organisms grow and stay healthy. Scientists study viruses to understand how they work and find ways to keep us, our pets, and our plants safe.

WOLBACHIA

(waal·baa·kee·uh)

Wolbachia are a group of rod-shaped bacteria that live inside insects, spiders, and other small creatures. These bacteria can help provide their hosts with nutrients. Other times, it tricks them into making more Wolbachia. Scientists are very interested in this because it might help control bugs that bother crops or spread sickness.

X EROPHILE

(zee·ro·fai·uhl)

Meet the xerophiles, tiny living things that love dry places! Xerophiles are special microorganisms that thrive in places with very little water, like deserts. Xerophiles teach scientists how life can adapt to extreme conditions, showing us the incredible ways living things can survive in places with little to no water.

Yeast

Yeast are tiny organisms that belong to the fungi family, the same family as a mushroom. Yeast, however, are very small and made of only one cell. They can be found almost everywhere in nature, like on fruits and flowers. In cooking, yeast helps make bread fluffy. They eat the sugar in the food and make a gas, called carbon dioxide, this is what makes the bread rise. You can find yeast in a bottle in your grocery store.

Z OONOSIS

(zo·on·o·ses)

Zoonosis is a big word that describes how diseases can pass from animals to people. In microbiology, scientists study tiny organisms like bacteria and viruses that can cause these diseases. Some examples of zoonosis include diseases like rabies, which can spread from animals like dogs to humans. Understanding zoonosis helps scientists and doctors keep people and animals healthy by knowing how diseases can move between them.

www.ingramcontent.com/pod-product-compliance
Lightning Source LLC
Chambersburg PA
CBHW081004290526

45795CB00009B/3080